curious about

BOMBERS

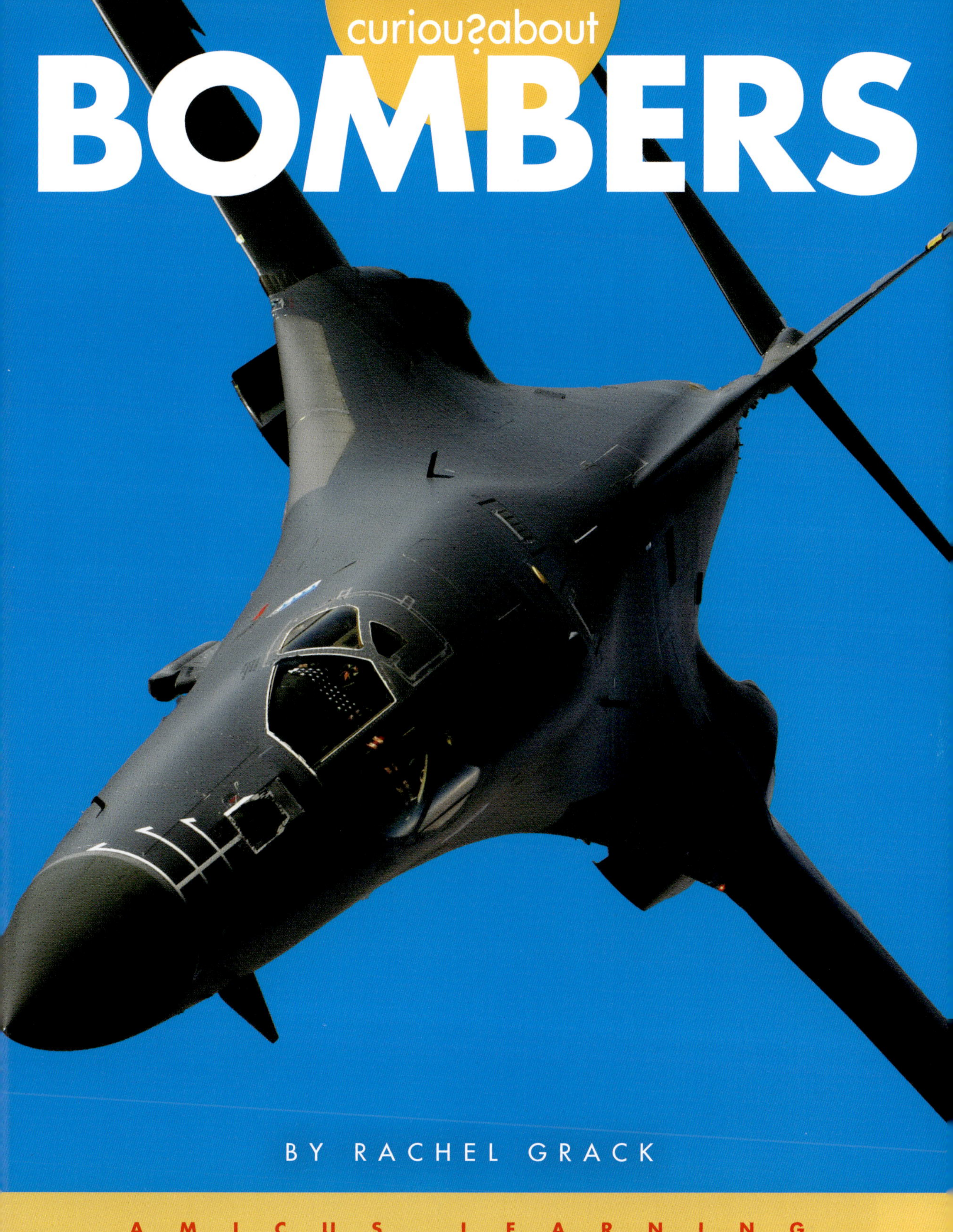

BY RACHEL GRACK

AMICUS LEARNING

What are you

curious about?

3

CHAPTER THREE

In the Skies

Curious About is published by
Amicus Learning, an imprint of Amicus
P.O. Box 227
Mankato, MN 56002
www.amicuspublishing.us

Editor: Alissa Thielges
Series Designer: Kathleen Petelinsek
Book Designer: Aubrey Harper

Library of Congress Cataloging-in-Publication Data
Names: Koestler-Grack, Rachel A., 1973– author.
Title: Curious about bombers / by Rachel Grack.
Description: Mankato, MN : Amicus Learning, [2025] | Series: Curious about military machines | Includes bibliographical references and index. | Audience: Ages 5–9 | Audience: Grades 2–3 | Summary: "Early elementary readers learn how bombers work in this inquiry-based nonfiction book about the airplanes' size, speed, and battle strength. Includes infographics and back matter to support research skills, plus table of contents, glossary, and index"—Provided by publisher.
Identifiers: LCCN 2023038614 (print) | LCCN 2023038615 (ebook) | ISBN 9781645493136 (library binding) | ISBN 9781645494010 (ebook)
Subjects: LCSH: Bombers—Juvenile literature. | Airplanes, Military—Juvenile literature.
Classification: LCC UG1242.B6 K644 2025 (print) | LCC UG1242.B6 (ebook) | DDC 623.74/63—dc23/eng/20231107
LC record available at https://lccn.loc.gov/2023038614
LC ebook record available at https://lccn.loc.gov/2023038615

Photo credits: DVIDS/Airman 1st Class Christopher Quail 6, Master Sgt. Matthew Plew 14–15, Master Sgt. William Greer cover, 1, 5, Petty Officer 3rd Class Dan Serianni 12, Senior Airman Austin McIntosh 20, Senior Airman Dwane Young 8–9, Senior Airman Jeffrey Withrow 13 (B-1), Senior Airman Jerreht Harris 21, Staff Sgt. Tiffany Emery 13 (B-52), Tech. Sgt. Corban Lundborg 16–17, Tech. Sgt. Heather Salazar 10–11; iStock/Jiri Podpinka 9 (icon), Yevhenii Dubinko 9 (football field); Northrop Grumman 18–19; Noun Project/ Supanut Piyakanont 22 & 23 (icons); Wikimedia Commons/ Dmitry Terekhov 7, Senior Airman Joel Pfiester 13 (B-2)

Printed in China

What are bombers?

Bombers are warplanes that bomb enemy **targets**. They also shoot missiles and torpedoes. Some bombers are long-range planes. They strike buildings and roadways from far away. Other bombers fly shorter distances. They zoom down to strike troops and bases.

A B-1 Lancer flies on mission in Afghanistan.

DID YOU KNOW?
Planes were first used as bombers in 1911. Soldiers dropped hand grenades.

Are they fast?

Yes! Some are **supersonic**. The fastest bomber is the Tu-160 Blackjack. It is a Russian bomber. These bombers can reach 1,380 miles (2,200 kilometers) per hour. That's more than two times the speed of sound! The fastest U.S. bomber is the B-1 Lancer. Both are heavily **armed**.

B-1 LANCER "THE BONE"

Wingspan: 137 feet (42 m)
Speed: 945 miles (1,521 km) per hour
Range: 7,456 miles (12,000 km)
Cost: $350 million

The Tu-160 is also called "White Swan" in Russia.

Tu-160
Wingspan: 183 feet (56 m)
Speed: 1,380 miles (2,200 km) per hour
Range: 7,456 miles (12,000 km)
Cost: $270 million

How many weapons can a bomber carry?

Each type carries different amounts. Its **payload** is measured in weight. Heavy bombers carry the biggest loads. The payload of a B-52 is 70,000 pounds (31,500 kg). Light bombers are smaller and make quick hits. They carry lighter loads.

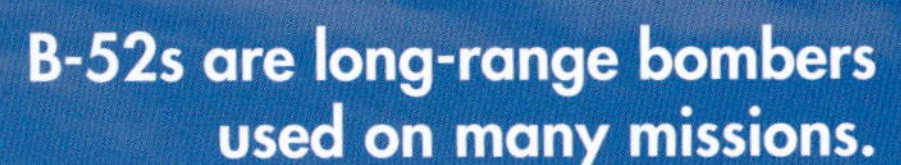
B-52s are long-range bombers used on many missions.

DID YOU KNOW?

The B-52 Stratofortress is the largest bomber in the world. It would cover half a football field.

CHAPTER TWO

What kind of fuel do bombers use?

A crew refuels a B-2 Spirit before takeoff.

They use JP-8. This is high-grade jet **fuel**. It can be stored longer than other types. It is good for both hot and cold weather. JP-8 also has a special **chemical** added to it. This helps it burn cleaner.

DID YOU KNOW?

Most military jet fuels have a high flash point. This means it takes more heat to set them on fire.

How many bombers are there?

The B-2 Spirit can attack at super high heights.

The U.S. Air Force has 152 bombers. That's more than any other country. Today, the Air Force uses three kinds of bombers. The B-1 and the B-52 are two of them. The third is the B-2 Spirit. It is a **stealth** bomber. This plane is invisible to enemy technology.

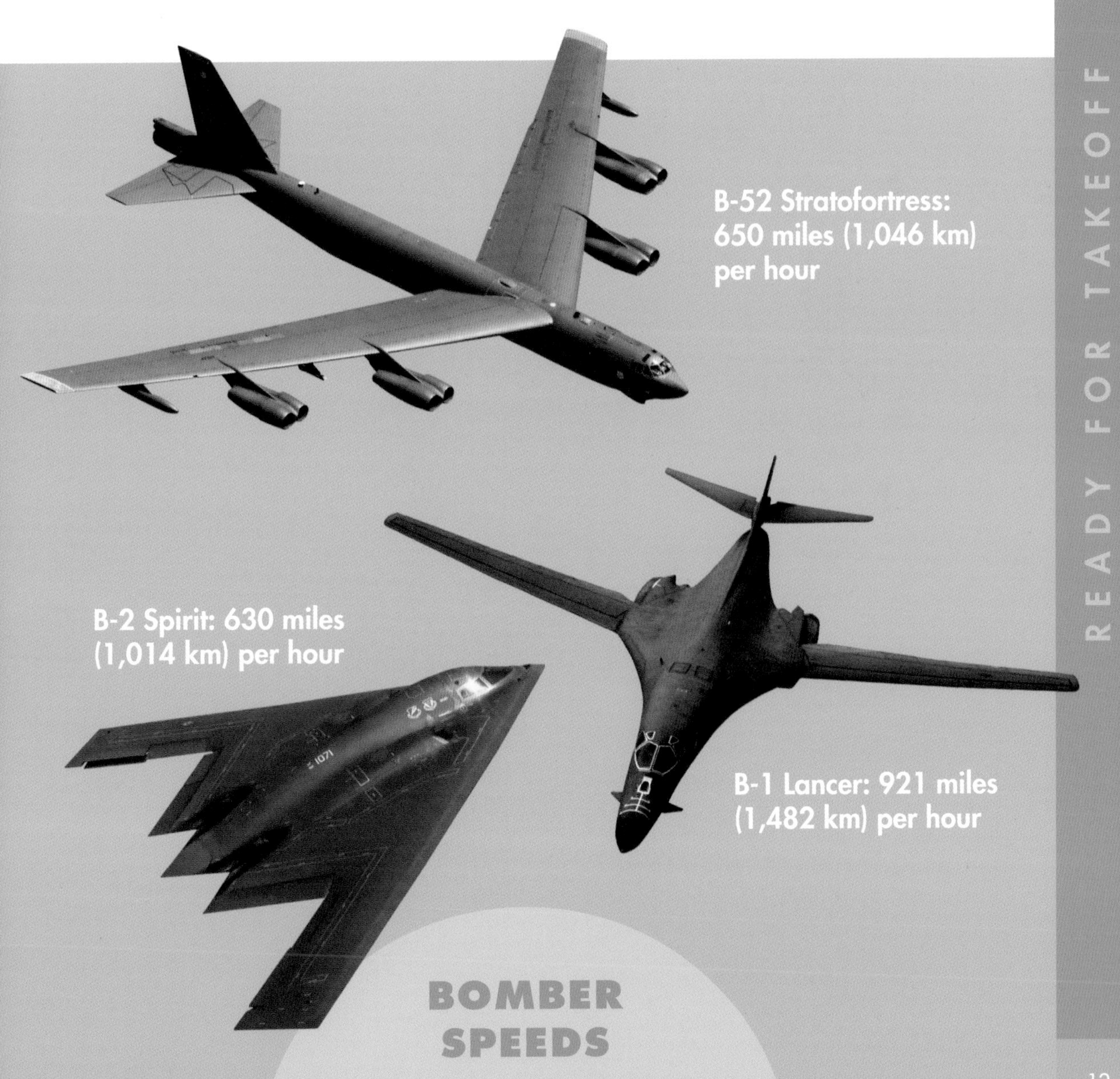

3

Why do B-2s have a weird shape?

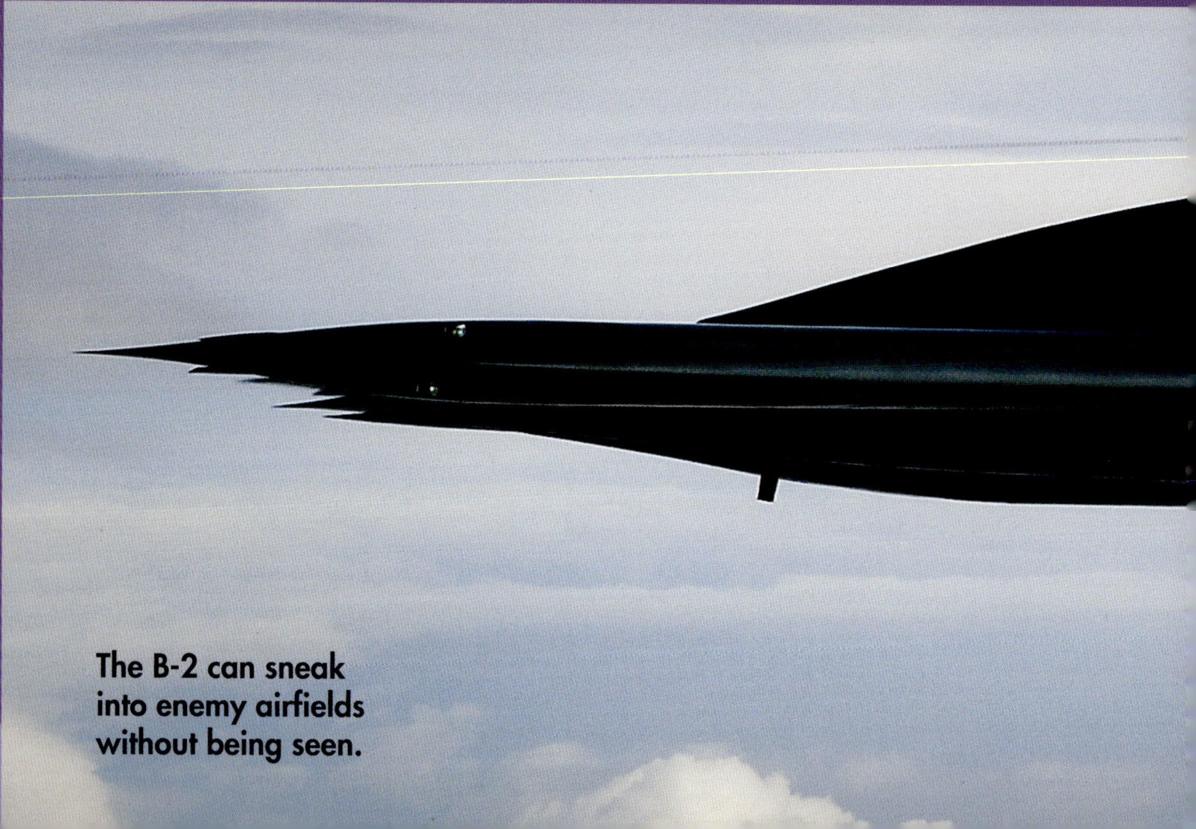

The B-2 can sneak into enemy airfields without being seen.

It helps hide them. Stealth bombers sneak up on the enemy. B-2s are flat and black. They disappear in the night sky. Their weird shape tricks enemies. It is hard to tell which direction they are going. It is also **aerodynamic**. It lets them fly farther on less fuel.

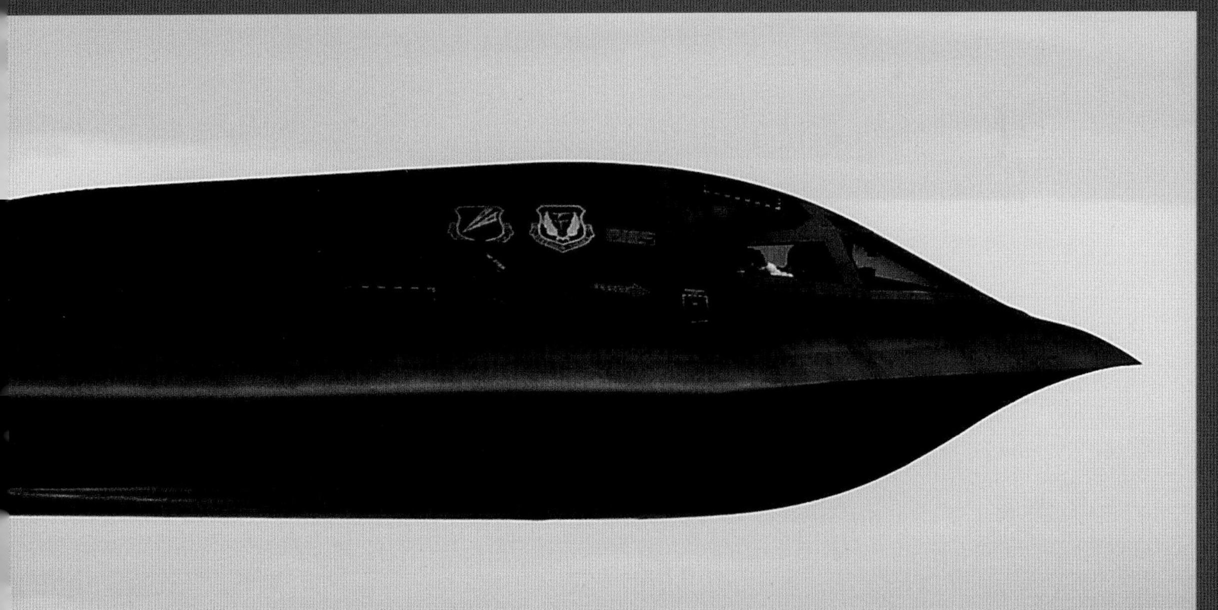

Two B-52 pilots keep a close eye on all the controls during takeoff.

What's it like to fly a bomber?

DID YOU KNOW?
The longest bomber mission was over 73 hours long. A B-2 flew from the United States to the Middle East and back again.

Hard and tiring. The cockpit is packed with computers, pedals, and levers. There are lots of buttons and lights. And you must master them all. You often fly all day or more. That's a long time in a small space. At least cockpits have a toilet!

How big is the flight crew?

The new B-21 will someday replace the B-1 and B-2 bombers.

That depends on the plane. The B-2 has a crew of two Air Force pilots. But the B-1 has a crew of four. Two fly the plane. The other two control the weapons. B-52s carry five crew members. An extra **navigator** watches their course.

DID YOU KNOW?

A new U.S. bomber took flight in 2023. The B-21 Raider can fly without a human crew. It uses artificial intelligence.

How do bombers refuel on long flights?

A B-1 gets into position to refuel.

In the air! The bomber meets up with a tanker plane. The bomber flies beneath the tanker. A hose drops down. It hooks up to the bomber. The planes fly together until the tank is full. It takes about 15 minutes. Then it is ready to roar!

The fuel hose attaches to the B-1's nose.

DID YOU KNOW?
B-2s can fly 6,200 miles (9,978 km) on a full tank. They need up to four air refills per operation.

ASK MORE QUESTIONS

What was the first bomber plane?

Are there any famous bomber pilots?

Try a BIG QUESTION: What skills are important for a bomber pilot?

SEARCH FOR ANSWERS

Search the library catalog or the Internet.
A librarian, teacher, or parent can help you.

Using Keywords
Find the looking glass.

Keywords are the most important words in your question.

?

If you want to know about:

- when the first bomber plane was built, type: FIRST BOMBER
- some famous pilots, type: FAMOUS BOMBER PILOTS

FIND GOOD SOURCES

Here are some good, safe sources you can use in your research.
Your librarian can help you find more.

Books

B-1B Lancer
by Megan Cooley Peterson, 2019.

Powerful Military Aircraft
by Cynthia Kennedy Henzel, 2023.

Internet Sites

DK Find Out!: U.S. Bomber Plane
www.dkfindout.com/us/history/world-war-ii/us-bomber-plane/
DK Find Out! is a website for kids. Discover educational topics through interactive pictures and text.

Kiddle: Bomber Facts for Kids
https://kids.kiddle.co/Bomber
Kiddle is an online encyclopedia for kids. Search this educational site for information on almost any topic!

Every effort has been made to ensure that these websites are appropriate for children. However, because of the nature of the Internet, it is impossible to guarantee that these sites will remain active indefinitely or that their contents will not be altered.

SHARE AND TAKE ACTION

Design a bomber.
Ask an adult to help you research stealth aircraft. Draw the shape you think would be hardest to see.

Test payloads.
Make two paper planes, a small one and a large one. Attach paper clips to each one-by-one. Which plane can carry the largest payload?

Talk it over with a friend.
Which is more important: stealth or speed? Why?

GLOSSARY

aerodynamic Something that is shaped in a way that helps it move easily through air.

armed Carrying weapons.

artificial intelligence AI for short; the power of a machine to copy intelligent human thinking and behaviors.

chemical A substance that can cause a change in another substance.

fuel A material, such as gas, that is burned to produce heat or power.

navigator A person who directs the course of an aircraft, ship, or other vehicle.

payload The amount of goods or materials that is carried by a vehicle.

stealth A secret and quiet way of moving or behaving.

supersonic A speed greater than that of sound.

target An object that something shoots at.

INDEX

About the Author

Rachel Grack has been editing and writing children's books since 1999. She lives on a small ranch in Arizona. Rachel holds deep respect for the U.S. military. Her uncle served as a first lieutenant in the Marines. He was the pilot of an F-8 Crusader.